This planner belongs to :

_____

2025

January
S	M	T	W	T	F	S
			1	2	3	4
5	6	7	8	9	10	11
12	13	14	15	16	17	18
19	20	21	22	23	24	25
26	27	28	29	30	31	

February
S	M	T	W	T	F	S
						1
2	3	4	5	6	7	8
9	10	11	12	13	14	15
16	17	18	19	20	21	22
23	24	25	26	27	28	

March
S	M	T	W	T	F	S
						1
2	3	4	5	6	7	8
9	10	11	12	13	14	15
16	17	18	19	20	21	22
23	24	25	26	27	28	29
30	31					

April
S	M	T	W	T	F	S
		1	2	3	4	5
6	7	8	9	10	11	12
13	14	15	16	17	18	19
20	21	22	23	24	25	26
27	28	29	30			

May
S	M	T	W	T	F	S
				1	2	3
4	5	6	7	8	9	10
11	12	13	14	15	16	17
18	19	20	21	22	23	24
25	26	27	28	29	30	31

June
S	M	T	W	T	F	S
1	2	3	4	5	6	7
8	9	10	11	12	13	14
15	16	17	18	19	20	21
22	23	24	25	26	27	28
29	30					

July
S	M	T	W	T	F	S
		1	2	3	4	5
6	7	8	9	10	11	12
13	14	15	16	17	18	19
20	21	22	23	24	25	26
27	28	29	30	31		

August
S	M	T	W	T	F	S
					1	2
3	4	5	6	7	8	9
10	11	12	13	14	15	16
17	18	19	20	21	22	23
24	25	26	27	28	29	30
31						

September
S	M	T	W	T	F	S
	1	2	3	4	5	6
7	8	9	10	11	12	13
14	15	16	17	18	19	20
21	22	23	24	25	26	27
28	29	30				

October
S	M	T	W	T	F	S
			1	2	3	4
5	6	7	8	9	10	11
12	13	14	15	16	17	18
19	20	21	22	23	24	25
26	27	28	29	30	31	

November
S	M	T	W	T	F	S
						1
2	3	4	5	6	7	8
9	10	11	12	13	14	15
16	17	18	19	20	21	22
23	24	25	26	27	28	29
30						

December
S	M	T	W	T	F	S
	1	2	3	4	5	6
7	8	9	10	11	12	13
14	15	16	17	18	19	20
21	22	23	24	25	26	27
28	29	30	31			

Year in Pixels

	J	F	M	A	M	J	J	A	S	O	N	D
1.												
2.												
3.												
4.												
5.												
6.												
7.												
8.												
9.												
10.												
11.												
12.												
13.												
14.												
15.												
16.												
17.												
18.												
19.												
20.												
21.												
22.												
23.												
24.												
25.												
26.												
27.												
28.												
29.												
30.												
31.												

Color Codes

Notes

January 2025

MONDAY	TUESDAY	WEDNESDAY	THURSDAY
		1	2
6	7	8	9
13	14	15	16
20	21	22	23
27	28	29	30

January 2025

FRIDAY	SATURDAY	SUNDAY	NOTES
3	4	5	○
			○
			○
			○
			○
10	11	12	○
			○
			○
			○
17	18	19	○
			○
			○
			○
			○
24	25	26	○
			○
			○
			○
			○
31			NOTES

February 2025

MONDAY	TUESDAY	WEDNESDAY	THURSDAY
3	4	5	6
10	11	12	13
17	18	19	20
24	25	26	27

February

FRIDAY	SATURDAY	SUNDAY	NOTES
	1	2	○
			○
			○
			○
7	8	9	○
			○
			○
			○
14	15	16	○
			○
			○
			○
			○
21	22	23	○
			○
			○
			○
			○
28			NOTES

March 2025

MONDAY	TUESDAY	WEDNESDAY	THURSDAY
		1	2
6	7	8	9
13	14	15	16
20	21	22	23
27	28	29	30

March 2025

FRIDAY	SATURDAY	SUNDAY	NOTES
3	4	5	○
			○
			○
			○
			○
10	11	12	○
			○
			○
			○
			○
17	18	19	○
			○
			○
			○
24	25	26	○
			○
			○
			○
			○
31			NOTES

April

MONDAY	TUESDAY	WEDNESDAY	THURSDAY
	1	2	3
7	8	9	10
14	15	16	17
21	22	23	24
28	29	30	

April

FRIDAY	SATURDAY	SUNDAY	NOTES
4	5	6	○
			○
			○
			○
			○
11	12	13	○
			○
			○
			○
			○
18	19	20	○
			○
			○
			○
			○
25	26	27	○
			○
			○
			○
			○
			NOTES

May 2025

MONDAY	TUESDAY	WEDNESDAY	THURSDAY
			1
5	6	7	8
12	13	14	15
19	20	21	22
26	27	28	29

May 2025

FRIDAY	SATURDAY	SUNDAY	NOTES
2	3	4	○
			○
			○
			○
			○
9	10	11	○
			○
			○
			○
			○
16	17	18	○
			○
			○
			○
23	24	25	○
			○
			○
			○
			○
30	31		NOTES

June 2025

MONDAY	TUESDAY	WEDNESDAY	THURSDAY
2	3	4	5
9	10	11	12
16	17	18	19
23	24	25	26

June 2025

FRIDAY	SATURDAY	SUNDAY	NOTES
		1	○
			○
			○
			○
			○
6	7	8	○
			○
			○
			○
			○
13	14	15	○
			○
			○
			○
			○
20	21	22	○
			○
			○
			○
			○
27	28	29	30

July 2025

MONDAY	TUESDAY	WEDNESDAY	THURSDAY
	1	2	3
7	8	9	10
14	15	16	17
21	22	23	24
28	29	30	31

July 2025

FRIDAY	SATURDAY	SUNDAY	NOTES
4	5	6	○
			○
			○
			○
			○
11	12	13	○
			○
			○
			○
18	19	20	○
			○
			○
			○
			○
25	26	27	○
			○
			○
			○
			○
			NOTES

August 2025

MONDAY	TUESDAY	WEDNESDAY	THURSDAY
4	5	6	7
11	12	13	14
18	19	20	21
25	26	27	28

August

2025

FRIDAY	SATURDAY	SUNDAY	NOTES
1	2	3	○
			○
			○
			○
			○
8	9	10	○
			○
			○
			○
			○
15	16	17	○
			○
			○
			○
			○
22	23	24	NOTES
29	30	31	

September 2025

MONDAY	TUESDAY	WEDNESDAY	THURSDAY
1	2	3	4
8	9	10	11
15	16	17	18
22	23	24	25
29	30		

September 2025

FRIDAY	SATURDAY	SUNDAY	NOTES
5	6	7	○
			○
			○
			○
			○
12	13	14	○
			○
			○
			○
			○
19	20	21	○
			○
			○
			○
26	27	28	○
			○
			○
			○
			○
			NOTES

October 2025

MONDAY	TUESDAY	WEDNESDAY	THURSDAY
		1	2
6	7	8	9
13	14	15	16
20	21	22	23
27	28	29	30

October  2025

FRIDAY	SATURDAY	SUNDAY	NOTES
3	4	5	○
			○
			○
			○
			○
10	11	12	○
			○
			○
			○
			○
17	18	19	○
			○
			○
			○
			○
24	25	26	○
			○
			○
			○
			○
31			NOTES

November

MONDAY	TUESDAY	WEDNESDAY	THURSDAY
3	4	5	6
10	11	12	13
17	18	19	20
24	25	26	27

November 2025

FRIDAY	SATURDAY	SUNDAY	NOTES
	1	2	○
			○
			○
			○
			○
7	8	9	○
			○
			○
			○
14	15	16	○
			○
			○
			○
			○
21	22	23	○
			○
			○
			○
			○
28	29	30	NOTES

December

THURSDAY	TUESDAY	WEDNESDAY	THURSDAY
1	2	3	4
8	9	10	11
15	16	17	18
22	23	24	25
29	30	31	

December

2025

FRIDAY	SATURDAY	SUNDAY	NOTES
5	6	7	○
			○
			○
			○
			○
12	13	14	○
			○
			○
			○
			○
19	20	21	○
			○
			○
			○
26	27	28	○
			○
			○
			○
			○
			NOTES

December
2024

01 SUNDAY

02 MONDAY

03 TUESDAY

04 WEDNESDAY

December
2024

05 THURSDAY

- ○
- ○
- ○
- ○
- ○
- ○
- ○
- ○

06 FRIDAY

- ○
- ○
- ○
- ○
- ○
- ○
- ○
- ○

07 SATURDAY

- ○
- ○
- ○
- ○
- ○
- ○
- ○
- ○

08 SUNDAY

- ○
- ○
- ○
- ○
- ○

09 MONDAY

10 TUESDAY

11 WEDNESDAY

12 THURSDAY

13 FRIDAY

○
○
○
○
○
○
○
○

14 SATURDAY

○
○
○
○
○
○
○
○

15 SUNDAY

○
○
○
○
○
○
○
○

16 MONDAY

○
○
○
○
○

17 TUESDAY

18 WEDNESDAY

19 THURSDAY

20 FRIDAY

21 SATURDAY
- ○
- ○
- ○
- ○
- ○
- ○
- ○
- ○

22 SUNDAY
- ○
- ○
- ○
- ○
- ○
- ○
- ○
- ○

23 MONDAY
- ○
- ○
- ○
- ○
- ○
- ○
- ○
- ○

24 TUESDAY
- ○
- ○
- ○
- ○
- ○

25 WEDNESDAY

26 THURSDAY

27 FRIDAY

28 SATURDAY

29 SUNDAY

- ○
- ○
- ○
- ○
- ○
- ○
- ○
- ○

30 MONDAY

- ○
- ○
- ○
- ○
- ○
- ○
- ○
- ○

31 TUESDAY

- ○
- ○
- ○
- ○
- ○
- ○
- ○
- ○

NOTES

January
2025

01 WEDNESDAY

02 THURSDAY

03 FRIDAY

04 SATURDAY

January
2025

05 SUNDAY
- ○
- ○
- ○
- ○
- ○
- ○
- ○
- ○

06 MONDAY
- ○
- ○
- ○
- ○
- ○
- ○
- ○
- ○

07 TUESDAY
- ○
- ○
- ○
- ○
- ○
- ○
- ○
- ○

08 WEDNESDAY
- ○
- ○
- ○
- ○
- ○

09 THURSDAY

○
○
○
○
○
○
○
○

10 FRIDAY

○
○
○
○
○
○
○
○

11 SATURDAY

○
○
○
○
○
○
○
○

12 SUNDAY

○
○
○
○
○

13 MONDAY

○ _____
○ _____
○ _____
○ _____
○ _____
○ _____
○ _____
○ _____

14 TUESDAY

○ _____
○ _____
○ _____
○ _____
○ _____
○ _____
○ _____
○ _____

15 WEDNESDAY

○ _____
○ _____
○ _____
○ _____
○ _____
○ _____
○ _____
○ _____

16 THURSDAY

○ _____
○ _____
○ _____
○ _____
○ _____

17 FRIDAY

18 SATURDAY

19 SUNDAY

20 MONDAY

21 TUESDAY

22 WEDNESDAY

23 THURSDAY

24 FRIDAY

25 SATURDAY

○ _____
○ _____
○ _____
○ _____
○ _____
○ _____
○ _____
○ _____

26 SUNDAY

○ _____
○ _____
○ _____
○ _____
○ _____
○ _____
○ _____
○ _____

27 MONDAY

○ _____
○ _____
○ _____
○ _____
○ _____
○ _____
○ _____
○ _____

28 TUESDAY

○ _____
○ _____
○ _____
○ _____
○ _____

29 WEDNESDAY
- ○
- ○
- ○
- ○
- ○
- ○
- ○
- ○

30 THURSDAY
- ○
- ○
- ○
- ○
- ○
- ○
- ○
- ○

31 FRIDAY
- ○
- ○
- ○
- ○
- ○
- ○
- ○
- ○

NOTES

February
2025

01 SATURDAY

○ _____
○ _____
○ _____
○ _____
○ _____
○ _____
○ _____
○ _____

02 SUNDAY

○ _____
○ _____
○ _____
○ _____
○ _____
○ _____
○ _____
○ _____

03 MONDAY

○ _____
○ _____
○ _____
○ _____
○ _____
○ _____
○ _____
○ _____

04 TUESDAY

○ _____
○ _____
○ _____
○ _____
○ _____

February 2025

05 WEDNESDAY

06 THURSDAY

07 FRIDAY

08 SATURDAY

February
2025

09 SUNDAY
○
○
○
○
○
○
○
○

10 MONDAY
○
○
○
○
○
○
○
○

11 TUESDAY
○
○
○
○
○
○
○
○

12 WEDNESDAY
○
○
○
○
○

13 THURSDAY

14 FRIDAY

15 SATURDAY

16 SUNDAY

February
2025

17 MONDAY

○
○
○
○
○
○
○
○

18 TUESDAY

○
○
○
○
○
○
○
○

19 WEDNESDAY

○
○
○
○
○
○
○
○

20 THURSDAY

○
○
○
○
○

February
2025

21 FRIDAY

22 SATURDAY

23 SUNDAY

24 MONDAY

25 TUESDAY

26 WEDNESDAY

27 THURSDAY

28 FRIDAY

01 SATURDAY
- ○
- ○
- ○
- ○
- ○
- ○
- ○
- ○

02 SUNDAY
- ○
- ○
- ○
- ○
- ○
- ○
- ○
- ○

03 MONDAY
- ○
- ○
- ○
- ○
- ○
- ○
- ○
- ○

04 TUESDAY
- ○
- ○
- ○
- ○
- ○

05 WEDNESDAY

06 THURSDAY

07 FRIDAY

08 SATURDAY

09 SUNDAY

10 MONDAY

11 TUESDAY

12 WEDNESDAY

March 2025

13 THURSDAY

14 FRIDAY

15 SATURDAY

16 SUNDAY

17 MONDAY

18 TUESDAY

19 WEDNESDAY

20 THURSDAY

March
2025

21 FRIDAY

22 SATURDAY

23 SUNDAY

24 MONDAY

25 TUESDAY

26 WEDNESDAY

27 THURSDAY

28 FRIDAY

29 SATURDAY

○ _____
○ _____
○ _____
○ _____
○ _____
○ _____
○ _____
○ _____

30 SUNDAY

○ _____
○ _____
○ _____
○ _____
○ _____
○ _____
○ _____
○ _____

31 MONDAY

○ _____
○ _____
○ _____
○ _____
○ _____
○ _____
○ _____
○ _____

NOTES

_____
_____
_____
_____

April
2025

01 TUESDAY

○
○
○
○
○
○
○
○

02 WEDNESDAY

○
○
○
○
○
○
○
○

03 THURSDAY

○
○
○
○
○
○
○
○

04 FRIDAY

○
○
○
○
○

05 SATURDAY

06 SUNDAY

07 MONDAY

08 TUESDAY

April
2025

09 WEDNESDAY

10 THURSDAY

11 FRIDAY

12 SATURDAY

13 SUNDAY

14 MONDAY

15 TUESDAY

16 WEDNESDAY

17 THURSDAY

○ _____
○ _____
○ _____
○ _____
○ _____
○ _____
○ _____
○ _____

18 FRIDAY

○ _____
○ _____
○ _____
○ _____
○ _____
○ _____
○ _____
○ _____

19 SATURDAY

○ _____
○ _____
○ _____
○ _____
○ _____
○ _____
○ _____
○ _____

20 SUNDAY

○ _____
○ _____
○ _____
○ _____
○ _____

21 MONDAY

22 TUESDAY

23 WEDNESDAY

24 THURSDAY

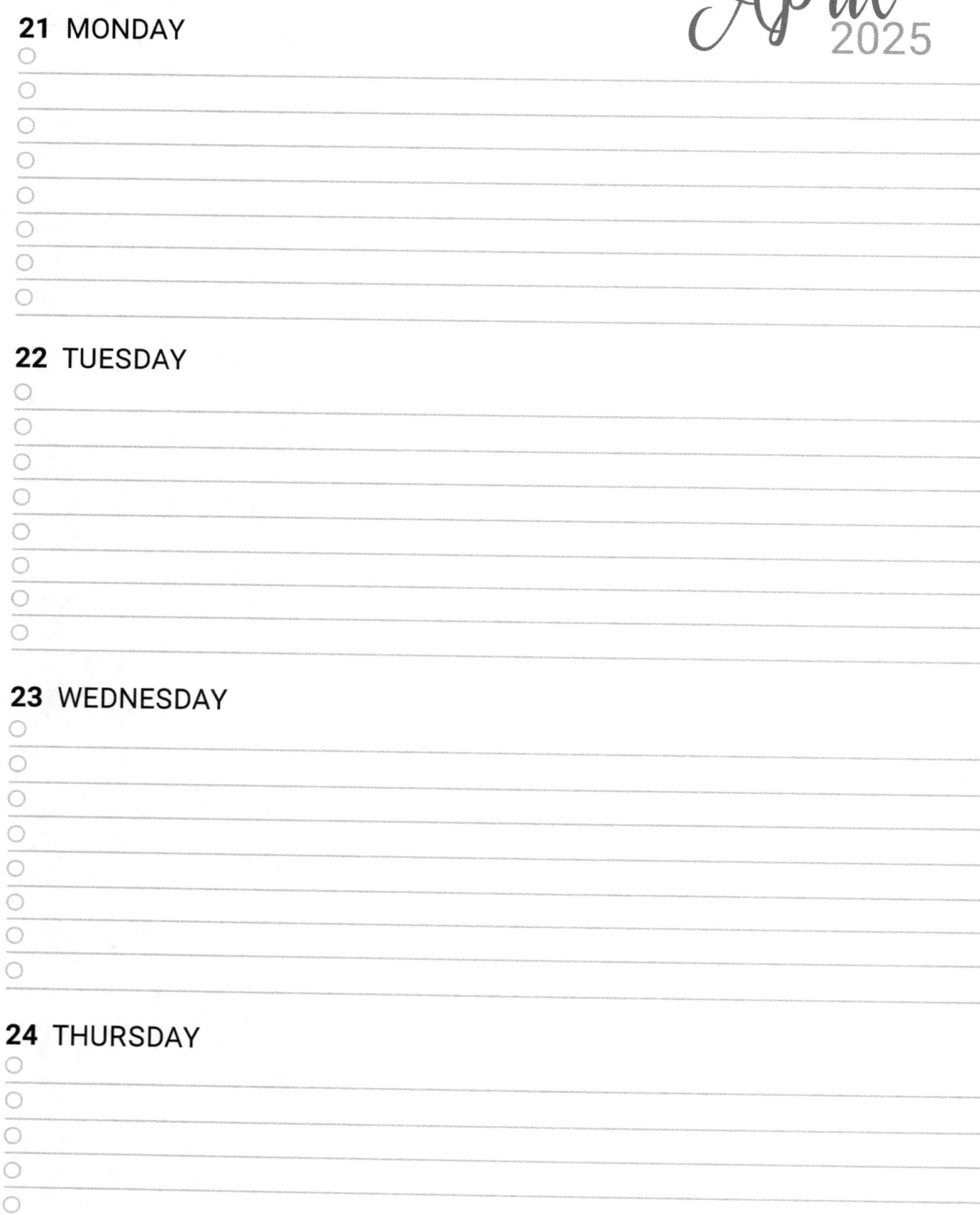

April
2025

25 FRIDAY

26 SATURDAY

27 SUNDAY

28 MONDAY

29 TUESDAY

30 WEDNESDAY

NOTES

01 THURSDAY

02 FRIDAY

03 SATURDAY

04 SUNDAY

May
2025

05 MONDAY

06 TUESDAY

07 WEDNESDAY

08 THURSDAY

09 FRIDAY

10 SATURDAY

11 SUNDAY

12 MONDAY

May
2025

13 TUESDAY

14 WEDNESDAY

15 THURSDAY

16 FRIDAY

17 SATURDAY

18 SUNDAY

19 MONDAY

20 TUESDAY

21 WEDNESDAY

○
○
○
○
○
○
○
○

22 THURSDAY

○
○
○
○
○
○
○
○

23 FRIDAY

○
○
○
○
○
○
○
○

24 SATURDAY

○
○
○
○
○

May
2025

25 SUNDAY

- ○ ....
- ○ ....
- ○ ....
- ○ ....
- ○ ....
- ○ ....
- ○ ....
- ○ ....

26 MONDAY

- ○ ....
- ○ ....
- ○ ....
- ○ ....
- ○ ....
- ○ ....
- ○ ....
- ○ ....

27 TUESDAY

- ○ ....
- ○ ....
- ○ ....
- ○ ....
- ○ ....
- ○ ....
- ○ ....
- ○ ....

28 WEDNESDAY

- ○ ....
- ○ ....
- ○ ....
- ○ ....
- ○ ....

May
2025

29 THURSDAY

30 FRIDAY

31 SATURDAY

NOTES

June
2025

01 SUNDAY

○ _____
○ _____
○ _____
○ _____
○ _____
○ _____
○ _____
○ _____

02 MONDAY

○ _____
○ _____
○ _____
○ _____
○ _____
○ _____
○ _____
○ _____

03 TUESDAY

○ _____
○ _____
○ _____
○ _____
○ _____
○ _____
○ _____
○ _____

04 WEDNESDAY

○ _____
○ _____
○ _____
○ _____
○ _____

05 THURSDAY

○ _____
○ _____
○ _____
○ _____
○ _____
○ _____
○ _____
○ _____

06 FRIDAY

○ _____
○ _____
○ _____
○ _____
○ _____
○ _____
○ _____
○ _____

07 SATURDAY

○ _____
○ _____
○ _____
○ _____
○ _____
○ _____
○ _____
○ _____

08 SUNDAY

○ _____
○ _____
○ _____
○ _____
○ _____

09 MONDAY

10 TUESDAY

11 WEDNESDAY

12 THURSDAY

June 2025

13 FRIDAY

14 SATURDAY

15 SUNDAY

16 MONDAY

17 TUESDAY

18 WEDNESDAY

19 THURSDAY

20 FRIDAY

June
2025

21 SATURDAY

22 SUNDAY

23 MONDAY

24 TUESDAY

25 WEDNESDAY

○ _____
○ _____
○ _____
○ _____
○ _____
○ _____
○ _____
○ _____

26 THURSDAY

○ _____
○ _____
○ _____
○ _____
○ _____
○ _____
○ _____
○ _____

27 FRIDAY

○ _____
○ _____
○ _____
○ _____
○ _____
○ _____
○ _____
○ _____

28 SATURDAY

○ _____
○ _____
○ _____
○ _____
○ _____

29 SUNDAY

30 MONDAY

NOTES

July
2025

01 TUESDAY
- ○ _____
- ○ _____
- ○ _____
- ○ _____
- ○ _____
- ○ _____
- ○ _____
- ○ _____

02 WEDNESDAY
- ○ _____
- ○ _____
- ○ _____
- ○ _____
- ○ _____
- ○ _____
- ○ _____
- ○ _____

03 THURSDAY
- ○ _____
- ○ _____
- ○ _____
- ○ _____
- ○ _____
- ○ _____
- ○ _____
- ○ _____

04 FRIDAY
- ○ _____
- ○ _____
- ○ _____
- ○ _____
- ○ _____

05 SATURDAY

○
○
○
○
○
○
○
○

06 SUNDAY

○
○
○
○
○
○
○
○

07 MONDAY

○
○
○
○
○
○
○
○

08 TUESDAY

○
○
○
○

09 WEDNESDAY

10 THURSDAY

11 FRIDAY

12 SATURDAY

13 SUNDAY

14 MONDAY

15 TUESDAY

16 WEDNESDAY

17 THURSDAY
- ○
- ○
- ○
- ○
- ○
- ○
- ○
- ○

18 FRIDAY
- ○
- ○
- ○
- ○
- ○
- ○
- ○
- ○

19 SATURDAY
- ○
- ○
- ○
- ○
- ○
- ○
- ○
- ○

20 SUNDAY
- ○
- ○
- ○
- ○
- ○

21 MONDAY
○
○
○
○
○
○
○
○

22 TUESDAY
○
○
○
○
○
○
○
○

23 WEDNESDAY
○
○
○
○
○
○
○
○

24 THURSDAY
○
○
○
○
○

25 FRIDAY
- ○
- ○
- ○
- ○
- ○
- ○
- ○
- ○

26 SATURDAY
- ○
- ○
- ○
- ○
- ○
- ○
- ○
- ○

27 SUNDAY
- ○
- ○
- ○
- ○
- ○
- ○
- ○
- ○

28 MONDAY
- ○
- ○
- ○
- ○
- ○

29 TUESDAY

30 WEDNESDAY

31 THURSDAY

NOTES

August
2025

01 FRIDAY

○ _____
○ _____
○ _____
○ _____
○ _____
○ _____
○ _____
○ _____

02 SATURDAY

○ _____
○ _____
○ _____
○ _____
○ _____
○ _____
○ _____
○ _____

03 SUNDAY

○ _____
○ _____
○ _____
○ _____
○ _____
○ _____
○ _____
○ _____

04 MONDAY

○ _____
○ _____
○ _____
○ _____
○ _____

05 TUESDAY

06 WEDNESDAY

07 THURSDAY

08 FRIDAY

09 SATURDAY

10 SUNDAY

11 MONDAY

12 TUESDAY

13 WEDNESDAY

14 THURSDAY

15 FRIDAY

16 SATURDAY

August
2025

17 SUNDAY

18 MONDAY

19 TUESDAY

20 WEDNESDAY

21 THURSDAY

22 FRIDAY

23 SATURDAY

24 SUNDAY

25 MONDAY

○ _____
○ _____
○ _____
○ _____
○ _____
○ _____
○ _____
○ _____

26 TUESDAY

○ _____
○ _____
○ _____
○ _____
○ _____
○ _____
○ _____
○ _____
○ _____

27 WEDNESDAY

○ _____
○ _____
○ _____
○ _____
○ _____
○ _____
○ _____
○ _____

28 THURSDAY

○ _____
○ _____
○ _____
○ _____
○ _____

29 FRIDAY

○ _____
○ _____
○ _____
○ _____
○ _____
○ _____
○ _____
○ _____

30 SATURDAY

○ _____
○ _____
○ _____
○ _____
○ _____
○ _____
○ _____
○ _____

31 SUNDAY

○ _____
○ _____
○ _____
○ _____
○ _____
○ _____
○ _____
○ _____

NOTES

_____
_____
_____
_____

01 MONDAY

○
○
○
○
○
○
○
○

02 TUESDAY

○
○
○
○
○
○
○
○

03 WEDNESDAY

○
○
○
○
○
○
○
○

04 THURSDAY

○
○
○
○
○

05 FRIDAY

○ _____
○ _____
○ _____
○ _____
○ _____
○ _____
○ _____
○ _____

06 SATURDAY

○ _____
○ _____
○ _____
○ _____
○ _____
○ _____
○ _____
○ _____

07 SUNDAY

○ _____
○ _____
○ _____
○ _____
○ _____
○ _____
○ _____
○ _____

08 MONDAY

○ _____
○ _____
○ _____
○ _____
○ _____

09 TUESDAY

10 WEDNESDAY

11 THURSDAY

12 FRIDAY

13 SATURDAY

○ _____
○ _____
○ _____
○ _____
○ _____
○ _____
○ _____
○ _____

14 SUNDAY

○ _____
○ _____
○ _____
○ _____
○ _____
○ _____
○ _____
○ _____

15 MONDAY

○ _____
○ _____
○ _____
○ _____
○ _____
○ _____
○ _____
○ _____

16 TUESDAY

○ _____
○ _____
○ _____
○ _____

17 WEDNESDAY

18 THURSDAY

19 FRIDAY

20 SATURDAY

21 SUNDAY

22 MONDAY

23 TUESDAY

24 WEDNESDAY

25 THURSDAY

26 FRIDAY

27 SATURDAY

28 SUNDAY

29 MONDAY

○ _____
○ _____
○ _____
○ _____
○ _____
○ _____
○ _____
○ _____

30 TUESDAY

○ _____
○ _____
○ _____
○ _____
○ _____
○ _____
○ _____
○ _____

NOTES

_____
_____
_____
_____
_____
_____
_____
_____
_____
_____
_____
_____
_____
_____

October
2025

01 WEDNESDAY

02 THURSDAY

03 FRIDAY

04 SATURDAY

05 SUNDAY

06 MONDAY

07 TUESDAY

08 WEDNESDAY

09 THURSDAY

○ _____
○ _____
○ _____
○ _____
○ _____
○ _____
○ _____
○ _____

10 FRIDAY

○ _____
○ _____
○ _____
○ _____
○ _____
○ _____
○ _____
○ _____

11 SATURDAY

○ _____
○ _____
○ _____
○ _____
○ _____
○ _____
○ _____
○ _____

12 SUNDAY

○ _____
○ _____
○ _____
○ _____
○ _____

13 MONDAY

○ _____
○ _____
○ _____
○ _____
○ _____
○ _____
○ _____
○ _____

14 TUESDAY

○ _____
○ _____
○ _____
○ _____
○ _____
○ _____
○ _____
○ _____

15 WEDNESDAY

○ _____
○ _____
○ _____
○ _____
○ _____
○ _____
○ _____
○ _____

16 THURSDAY

○ _____
○ _____
○ _____
○ _____

17 FRIDAY

18 SATURDAY

19 SUNDAY

20 MONDAY

21 TUESDAY

22 WEDNESDAY

23 THURSDAY

24 FRIDAY

25 SATURDAY

26 SUNDAY

27 MONDAY

28 TUESDAY

October
2025

29 WEDNESDAY

30 THURSDAY

31 FRIDAY

NOTES

November
2025

01 SATURDAY
- ○ _____
- ○ _____
- ○ _____
- ○ _____
- ○ _____
- ○ _____
- ○ _____
- ○ _____

02 SUNDAY
- ○ _____
- ○ _____
- ○ _____
- ○ _____
- ○ _____
- ○ _____
- ○ _____
- ○ _____

03 MONDAY
- ○ _____
- ○ _____
- ○ _____
- ○ _____
- ○ _____
- ○ _____
- ○ _____
- ○ _____

04 TUESDAY
- ○ _____
- ○ _____
- ○ _____
- ○ _____
- ○ _____

05 WEDNESDAY

06 THURSDAY

07 FRIDAY

08 SATURDAY

November
2025

09 SUNDAY

- ○
- ○
- ○
- ○
- ○
- ○
- ○
- ○

10 MONDAY

- ○
- ○
- ○
- ○
- ○
- ○
- ○
- ○

11 TUESDAY

- ○
- ○
- ○
- ○
- ○
- ○
- ○
- ○

12 WEDNESDAY

- ○
- ○
- ○
- ○
- ○

13 THURSDAY

14 FRIDAY

15 SATURDAY

16 SUNDAY

November
2025

17 MONDAY

18 TUESDAY

19 WEDNESDAY

20 THURSDAY

November
2025

21 FRIDAY

○
○
○
○
○
○
○
○

22 SATURDAY

○
○
○
○
○
○
○
○

23 SUNDAY

○
○
○
○
○
○
○
○

24 MONDAY

○
○
○
○
○

25 TUESDAY

- ○ _____
- ○ _____
- ○ _____
- ○ _____
- ○ _____
- ○ _____
- ○ _____
- ○ _____

26 WEDNESDAY

- ○ _____
- ○ _____
- ○ _____
- ○ _____
- ○ _____
- ○ _____
- ○ _____
- ○ _____

27 THURSDAY

- ○ _____
- ○ _____
- ○ _____
- ○ _____
- ○ _____
- ○ _____
- ○ _____
- ○ _____

28 FRIDAY

- ○ _____
- ○ _____
- ○ _____
- ○ _____
- ○ _____

29 SATURDAY

○ _____
○ _____
○ _____
○ _____
○ _____
○ _____
○ _____
○ _____

30 SUNDAY

○ _____
○ _____
○ _____
○ _____
○ _____
○ _____
○ _____
○ _____

NOTES

_____
_____
_____
_____
_____
_____
_____
_____
_____
_____
_____
_____
_____

December
2025

01 MONDAY

○ _____
○ _____
○ _____
○ _____
○ _____
○ _____
○ _____
○ _____

02 TUESDAY

○ _____
○ _____
○ _____
○ _____
○ _____
○ _____
○ _____
○ _____

03 WEDNESDAY

○ _____
○ _____
○ _____
○ _____
○ _____
○ _____
○ _____
○ _____

04 THURSDAY

○ _____
○ _____
○ _____
○ _____
○ _____

December
2025

05 FRIDAY

06 SATURDAY

07 SUNDAY

08 MONDAY

09 TUESDAY

10 WEDNESDAY

11 THURSDAY

12 FRIDAY

13 SATURDAY

14 SUNDAY

15 MONDAY

16 TUESDAY

17 WEDNESDAY

○ _____
○ _____
○ _____
○ _____
○ _____
○ _____
○ _____
○ _____

18 THURSDAY

○ _____
○ _____
○ _____
○ _____
○ _____
○ _____
○ _____
○ _____

19 FRIDAY

○ _____
○ _____
○ _____
○ _____
○ _____
○ _____
○ _____
○ _____

20 SATURDAY

○ _____
○ _____
○ _____
○ _____
○ _____

21 SUNDAY

22 MONDAY

23 TUESDAY

24 WEDNESDAY

December
2025

25 THURSDAY

26 FRIDAY

27 SATURDAY

28 SUNDAY

December
2025

29 MONDAY

○ _____
○ _____
○ _____
○ _____
○ _____
○ _____
○ _____
○ _____

30 TUESDAY

○ _____
○ _____
○ _____
○ _____
○ _____
○ _____
○ _____
○ _____

31 WEDNESDAY

○ _____
○ _____
○ _____
○ _____
○ _____
○ _____
○ _____
○ _____

NOTES

_____
_____
_____
_____